Paulo Cesar Waidzik

# URBAN MOBILITY: a villain is the 1930 model CAR!

Paulo Cesar Waidzik

# URBAN MOBILITY: a villain is the 1930 model CAR!

In Sustainable Urban Mobility, all modes are important and must be harmonious with people's cities

**Imprint**

Any brand names and product names mentioned in this book are subject to trademark, brand or patent protection and are trademarks or registered trademarks of their respective holders. The use of brand names, product names, common names, trade names, product descriptions etc. even without a particular marking in this work is in no way to be construed to mean that such names may be regarded as unrestricted in respect of trademark and brand protection legislation and could thus be used by anyone.

Cover image: www.ingimage.com

This book is a translation from the original published under ISBN 978-613-9-69678-9.

Publisher:
Sciencia Scripts
is a trademark of
Dodo Books Indian Ocean Ltd. and OmniScriptum S.R.L publishing group

120 High Road, East Finchley, London, N2 9ED, United Kingdom
Str. Armeneasca 28/1, office 1, Chisinau MD-2012, Republic of Moldova, Europe
Printed at: see last page
**ISBN: 978-620-8-13633-8**

*"If I've been able to see further, it's because I've stood on the shoulders of giants."*

ISAAC NEWTON

I would like to dedicate this final project to my sons Davi and Lucas, who, at eight years old, still don't fully understand the reasons why we had to give up some of our walks and camping trips while we were working on this project.

Paulo Cesar Waidzik

# ACKNOWLEDGEMENTS

To GOD for the energy to overcome this stage with great courage.

To my children Davi, Lucas and Camila, who are the reason for my work.

To my dear wife Heloisa for her unconditional support and for giving up her time to carry out this task.

To Yumi, Ricardo and Osório for contributing the wealth and value of their experience and knowledge.

To the organisations and professionals who dedicate their efforts to carrying out field research and share their convictions and invaluable information.

To Positivo University for the opportunity to take part in this MBA programme.

To the course coordinator, Professor Cequinel, to the module lecturer, Professor Felice, to all the lecturers who, with great joy and commitment, provided us with much enrichment.

I would especially like to thank Professors Jamilton and Glávio for guiding me so hard so that we could reach the end of this achievement together.

And to all the MBA colleagues for the fantastic days and experiences we spent together on this endeavour.

Paulo Cesar Waidzik

WAIDZIK, Paulo Cesar[I]

MBA student in Strategic and Sustainable Energy Planning - 2014 Positivo University

**ABSTRACT -** The 21st century, with its megacities, where in Brazil 81.22% of the population is mainly urban, presents itself as the century of the urban planet and maximises the challenges of the car in Urban Mobility. The abundant generation of wealth and technological resources has led to an increase in consumption and easier access to the opportunities of modernity. In conurbations, mobility becomes a challenge due to the high number of journeys, space restrictions and the resulting lack of efficiency and productivity. In this scenario, public policies on urban mobility, with the aim of guaranteeing fluidity, have determined the privileging of collective means of transport to the detriment of individual motorised modes of transport, among which the car is considered a villain. This article attempts to identify, based on statistics and field data, the characteristics that have given the car this status, taking into account various aspects of efficiency and productivity. It finds that the car is an essential means of individual transport for urban mobility, but that its current characteristics do not harmonise with the demands of the urban environment, and that in this century it is no longer possible to use multifunctional mobility vehicles, and that vehicles can no longer meet urban and extra-urban needs simultaneously. The conclusion is that in cities things must be tailored to the human being and for the harmony of human beings and the planet, so that cars must be reinvented to contribute to sustainability and quality of life.

**Keywords:** Car. Efficiency in urban mobility. Urban mobility. Productivity in urban mobility. Quality of life in urban mobility.

---

[I] Alameda Dr. Carlos de Carvalho, 719, Apt. 901 - Batel - 80.430-180 - Curitiba - PR E-mail: paulo@motiva.net.br

# SUMMARY

5

# CHAPTER 1

INTRODUCTION

The 21st century is shaping up to be the century in which civilisation will have to direct a large part of its efforts towards reorganising cities, which were born without planning, grew with practically individual, independent initiatives and without a long-term vision, and many have become megacities with their various challenges surfacing today.

In 1930, economist John Keynes predicted that in a hundred years' time, humanity would face its permanent problem: how to use freedom from pressing economic concerns, how to occupy the leisure that science and economic gains would bring to live well, wisely and pleasantly? Now that we are less than 20 years away from the scenario proposed by Keynes, perhaps it would be appropriate to look at the great issue of the century: the urban planet. After all, if the 19th century was about empires and the 20th about nations, this one is about cities. And the immense innovations that are now being announced will take place in urban territory (LEITE, 2012, pg.04).

The rural exodus has progressively taken on more and more significant proportions. In 1920, only 10 per cent of the Brazilian population lived in urban areas, and fifty years later, in 1970, the percentage had reached 55.9 per cent. Today, according to data from the 2010 Demographic Census of the Brazilian Institute of Geography and Statistics (IBGE), almost one hundred and sixty-one million Brazilians live in urban areas, which corresponds to 81.22 per cent of the population. In just a few decades, Brazil has transformed from a predominantly rural country into a mostly urban one.

Throughout the history of civilisation, man has always had the desire and need to move over short and long distances, whether simply to explore new places or to search for the resources necessary for his survival and evolution.

This restlessness and need for human beings to move around is strongly present in cities, which are the modern environments where they live and spend most of their time doing their activities. Urban mobility can be defined as the ability of people and goods to move around the urban space in order to carry out their daily activities, such as work, catering, education, health, culture, recreation and leisure. If it is adequate, people and goods are able to carry out their activities in a time considered ideal, comfortably and safely.

And in this scenario of the complexity of cities, the sheer volume and concentration of people, the limited physical spaces, the various problems relating to sustainability and quality of life, when it comes to urban mobility where congestion is rampant, various authors, including current legislation, simply opt to prioritise non-motorised means of transport and public transport and practically condemn the car as the great villain.

It can be seen in item II of article 6 of Law 12.587 of 03/01/2012 that the National Urban Mobility Policy is guided, among other things, by the guideline that defines the *priority of non-motorised modes of transport over motorised ones and of public transport services over individual motorised transport.*

Furthermore, as stated in item I of article 15 of the Municipal Plan for Urban Mobility and Integrated Transport (PlanMob Curitiba) in the final proposal approved by CONCITIBA on 4 December 2008 - ON URBAN MOBILITY AND TRANSPORT, the municipal mobility policy is committed to facilitating travel and the movement of people and goods in the municipality, among other things, with the following general guideline: *prioritising collective transport over individual transport in road space.*

Also according to the position presented by renowned architect and urban planner Jaime Lerner on the subject of cars in his talk at the III SIBRT Best Practices in Latin America Congress, Lerner insisted on this point, stating that "there is no future for cities if they depend only on cars". Furthermore, according to Jaime Lerner, "inhaling car exhaust fumes is the new passive smoking". The visionary architect, urban planner and former mayor of Curitiba has long been clear about the growing social stigma surrounding the use of private vehicles. In fact, Lerner predicts that the private car will soon become as much of a social nuisance as cigarettes: "You can use it," he says, "but people will feel uncomfortable." Luís Gutiérrez, Strategic Director for Latin America at EMBARQ and Secretary General of SIBRT, summarised Lerner's thoughts in the following words: "The car is the cigarette of the future".

> Lerner in his book Acupuntura Urbana (Urban Acupuncture) on page 85 says that urban cholesterol is the accumulation in our veins and arteries of excessive car use. This affects the body and even people's minds. They soon begin to think that everything can be solved with the car. So they prepare the city just for the car. Viaducts, expressways... and car emissions.

It is understood that these laws and positions are motivated by the "1930 model AUTOMOBILE" that we will present later.

But is the car really the villain of urban mobility that it has been so exhaustively framed as, and that the solution would be to simply exclude the car from the modes and have the collective and non-motorised modes as sufficient for urban mobility?

By taking an X-ray of the car and its use in the urban environment, we can identify details of possible problems attributed to the car or possible injustices being committed to it, which has been present and coexisting in the world's cities for over a hundred years.

The aim of this article is to study urban mobility modes and in particular the car, which is a vehicle in the individual motorised mobility mode, to discover the reasons why the federal government, municipal government and urban planners are defining new policies and opinions that exclude the car from the modes, considering it a villain.

The scope of the study will be exclusively mobility in the urban context, with an assessment of individual motorised transport. Extra-urban aspects (see glossary), public transport and non-motorised transport will only be used as statistical data for comparative purposes. Freight mobility and the financial, cost and safety issues of any mode will not be addressed, nor will financial comparisons with other types of urban modes. Efficiency and productivity aspects will be considered (see glossary) as factors for the sustainability of cities and the planet and for people's quality of life. Basically, information from Brazil will be used, particularly from the city of São Paulo as it is the largest city in the country and from Curitiba as a representative of a large city. In the statistics and surveys produced or filtered, the physical space will be considered to be the last urban mile (see glossary) and the evaluation period will be working days from Monday to Friday between 7am and 7pm.

# CHAPTER 2

DATA OVERVIEW

In order to identify the car (see glossary), it is first necessary to contextualise the environment in which it will be used, the range of transport modes available and the profile of the mobility user.

## 2.1  The city

Cities, with their concentration of people, shops and services, are growing in volume, perimeter length and integration of centres. They swell with volumes concentrated in spaces that are becoming smaller, inducing more and more need to travel both in quantity and distance.

Technological development is making it easier for people to access wealth and goods with the consequent increase in consumption and justifying the increase in circulation.

There is a growing number of vehicles in Brazil. According to the National Traffic Department (Denatran), the vehicle fleet currently exceeds 83 million registrations, having doubled over the last ten years.

According to data from the Curitiba Urban Planning and Research Institute (IPPUC), the ratio of vehicles to population shows a significant increase in motorisation in the city of Curitiba. While in the 1990s this index showed that there were 2.67 people per vehicle, in 2012 this ratio was 1.36 people per vehicle, as shown in figure 1.

Figure 1: Population per vehicle since 1980. SOURCE: BPTRAN-PR / IPARDES / IBGE / IPPUC
Note: (*) In 1994 the BPTRAN recycled its archive system.

In the appendix at the end, there is a map of inhabitants per vehicle in Curitiba which shows the concentrated volume of vehicles in neighbourhoods, showing various nerve centres (see glossary) within the city, multiplying the existence of the last mile in the city. Figure 2 shows a projection study of the relationship between population and vehicle growth in Curitiba.

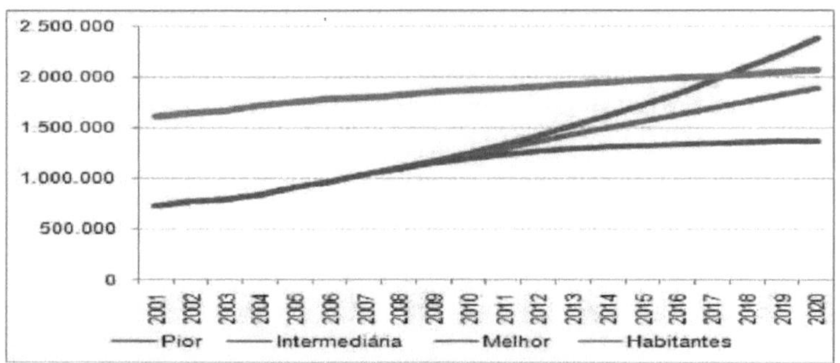


Figure 2: Population and vehicle projections for Curitiba (PR).
SOURCE: DETRAN/PR, IPPUC 2001 to 2008, IPPUC/Database 2009 to 2020 ELABORATION: Valéria Veríssimo Franco de Sousa - Statistics - 2010

From Monday to Friday, between 7am and 7pm, traffic jams on the roads are increasing. According to the 11 March 2014 edition of Bom Dia Brasil and published on the "gl.globo.com" website, São Paulo's traffic jams caused losses of R$40 billion in 2012, according to a calculation by the Getúlio Vargas Foundation. In five years, journeys around the city - including on foot - have grown by 15 per cent. In the same period, the population increased by just 2 per cent. The Getúlio Vargas Foundation did the calculation: the costs generated by traffic in São Paulo went from R$17 billion in 2002 to R$40 billion in 2012. Most of this was lost in productive hours. In ten years, this expense alone rose from R$10 billion to more than R$30 billion.

YAMAWAKI (2014) explains in his article "Smaller cars to help relieve traffic" that there are difficulties in expanding roads and car parks, which are physically limited to the demand, both in length and width of roads and car park areas.

There is an increasing volume of emissions (see glossary). The local pollutants considered are the following: CO (carbon monoxide), HC (hydrocarbons), $NO_x$ (Nitrogen Oxides), MP (particulate matter) and $SO_x$ (Sulphur Oxides), as defined by CETESB/SP. And CONAMA limits the maximum decibels allowed for vehicle emissions because there are emissions related to noise pollution generated by internal combustion motor vehicles.

## 2.2 Modes of urban mobility

The mode of transport used in this study is individual transport, specifically the car. However, seven other modes will be considered for the purposes of comparative parameters:

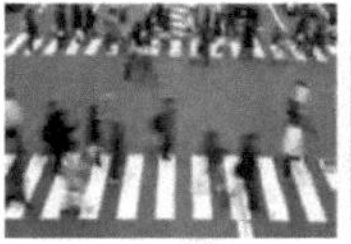

**A pé***          **Bicicleta**     **Moto e Automóvel***     **Ônibus e BRT***          **e-Trilhos***

The classification of modes defined in this work is:
**NMT:** Non-motorised transport (cycling and walking);
**IT:** Individual transport (car and motorbike);
**TC:** Public transport (buses, BRT, trains and metro).
*(see glossary)

## 2.3 The mobility user

The user is the person who uses the various modes to get around the city to carry out their daily activities. The CNI-IBOPE survey "Portraits of Brazilian Society: Urban Locomotion" of August 2011 provides information on how users use and evaluate the various modes of transport available and used by them, and reached the following conclusions, among others:

a) the vast majority of respondents consider their means of transport to be "excellent" or "good", especially in the case of individual means of transport. In the case of motorbikes and the family car, the combined percentages exceed 90% - 94% and 93% respectively - and the percentage of dissatisfied people is 2%;

b) the mode of transport that received the worst evaluation was the bus: 45 per cent of users are satisfied, or rate it as "excellent" or "good", but 24 per cent rate it as "bad" or "terrible";

c) the bus service is rated more favourably in small towns and in the interior of the states. Among residents of capital cities, the percentage who consider the service "excellent" or "good" is 36 per cent. In cities with up to 20,000 inhabitants, the percentage rises to 55 per cent;

d) is that 37% of the Brazilian population consider travelling time to be the main factor in choosing the mode of transport they use most;

e) when asked what makes them choose the means of transport they use, the most popular option was "It's the fastest" (37%). In second place, with 22 per cent, was the option "Only possibility/ No transport on offer";

f) in small towns with up to 20,000 inhabitants, the choice due to lack of availability was the most frequently mentioned (35 per cent). A similar situation occurs in the North/Centre-West: 29% chose their main means of transport because it was the only option and 25% because it was the fastest;

g) Among the positive aspects of the interviewee's main means of transport, the option "It's fast" was the most popular. This option was chosen by 54 per cent of those interviewed. The percentage rises to 87 per cent among motorbike users and falls to 33 per cent among those who use the bus as their main means of transport, although it remains the option most often ticked;

h) the option "Goes many places" came second, with 16 per cent of responses. The majority of respondents who chose this answer are "Minibus/ Van/ Beast/ Topic/ Kombi" users;

i) As for the time spent travelling, the study shows that in capitals and cities with more than 100,000 inhabitants, the longest time spent is between half an hour and two hours;

j) The main means of transport used to get around are buses, cars and walking, in that order;

l) in terms of the quality of the means of transport, the car comes first, followed by motorbikes, walking and cycling;

From this research it can be concluded that people prioritise comfort and quality in their mobility, and place great value on the time it takes to get around, and on the fact that the transport they choose can take care of all the activities they need to carry out.

# CHAPTER 3

CAR X-RAY

Just as general data was analysed in the previous chapter in order to identify the car, this chapter will survey and analyse data on the car in order to identify it.

## 3.1  Basic data on the car and its environment

The car is part of the individual motorised mode of urban mobility. It was invented, by the standards we know today, in the late 19th and early 20th centuries as a natural successor to the non-motorised cart that preceded it.

Today's car is the result of the historical choice that was made at that time. The car is defined in Annex I of the Brazilian Traffic Code as a vehicle for transporting up to eight passengers, excluding the driver.

The profile of the popular car available on the market basically has the following specifications and features: it occupies an area of approximately 8 m² (L=4m, W=2m), weighs around 1,000 kg, can carry five people and 290 litres of cargo volume; has a maximum final speed of between 140 km/h and 180 km/h, accelerates from 0 to 100 km/h in ten seconds, has a combustion engine and uses fossil fuels and lubricants; has a spare tyre.

The profile of traffic routes in cities has a maximum permitted speed of 60 km/h, if not 40 km/h. There are rare exceptions of up to 70 km/h. They have traffic lights and are approximately eight metres wide.

It can be seen that the average use of cars during their lifetime varies from 70 to 95 per cent in the city and from 5 to 30 per cent on the road.

## 3.2  The car as one of the modes of individual urban transport

Figure 3 shows various modes of urban mobility for people and how each one measures up in terms of some aspects of efficiency and productivity when used and applied in real day-to-day conditions and during periods considered to be "commercial" in cities, thus disregarding general technical information from manufacturers' laboratories.

The aim of the graph in figure 3 is to visualise and compare the modes according to their level of efficiency and productivity under each aspect defined in the analysis and to make it possible to visualise the particular performance of the car mode in relation to the others, thus making it easier to observe its strengths and weaknesses in the spectrum of urban mobility.

**3.2.1** Characteristics of the variables for the modal framework

Figure 3, and the following figures in this chapter, were generated taking into account the statistical data from the sources and the commonly understood considerations for each item in the legend:

a) for Land Use Efficiency (see glossary): The measurements of the approximate areas of the vehicles of each mode are considered in relation to the number of people it transports within the scope and parameters of the study, taking into account popular cars with large sales volumes in Brazil, standard buses, standard bi-articulated BRT and the 6-car underground metro; on foot, occupying an area of 0.25 $m^2$ ; bicycle, an area of 1.2 $m^2$ ; motorbike, an area of 1.6 $m^2$ ; car, an area of 8 $m^2$ ; bus, an area of 37.5 $m^2$ ; BRT, 65 $m^2$ ; metro, 409.5 $m^2$ ; these data were obtained from the websites of the various vehicle manufacturers in Brazil. On foot, one person; bicycle and motorbike, used by one person; car, used by one and a half people; bus, used by 110 people per journey; BRT, 230 people per journey; metro, 2000 people per journey (in 6 carriages). Information on car use can be found on the IPPUC website, buses on the Urbanização de Curitiba S.A. (URBS) website and the metro on the METRO/SP website listed in the references;

b) for Energy Efficiency (see glossary): considerations from the interview in the appendix, data from INEE and IPPUC;

c) for Efficiency in Gas and Noise Emissions (see glossary): ANTP - Urban Mobility Information System - General Report 2011 and CONAMA RESOLUTION $n^9$ 272;

d) for Productivity in the Last Mile and Beyond the Last Mile (see glossary): considers the agility in travelling to fulfil activities within the scope of the study, data from IPPUC and CNI (National Confederation of Industry);

e) for Productivity without load (see glossary): agility in travelling to fulfil activities within the scope of the study, without load;

f) **for Productivity with a small personal load** (see glossary): agility in travelling to carry out activities within the scope of the study, with a small **personal** load;

g) **for Efficiency in Transversal and Longitudinal Use** (see glossary): agility in making journeys to fulfil activities within the scope of the study in a transversal or longitudinal way, data from IPPUC and CNI;

The criterion for fitting the values in the graph in figure 3 was to consider three levels: low as the least efficient or least productive; medium; and high as the most efficient or most productive, by approximation or distance, within the spectrum shown in the statistical figures.

Purposely, from here on the *car* will be called *"AUTOMOBILE 2014/1930"* in reference to the "current AUTOMOBILE model 1930" and will be justified in the next chapter.

### 3.2.2 General comparison between the selected modes

Figure 3 shows the modes of urban mobility for people and their relationship with efficiency and productivity.

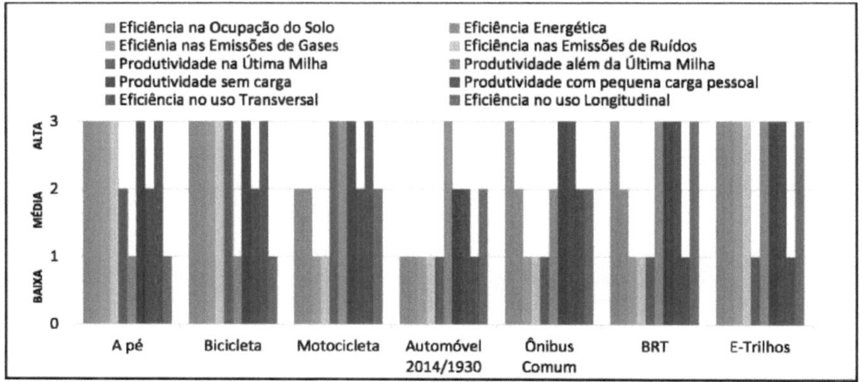


Figure 3: Urban mobility modes for people - efficiency and productivity. Source: The author (2014), considering the references in section 3.2.1.

Figure 4, a scatter plot based on the same information as figure 3, serves to make it easier to visualise the comparative trend of all modes on the same graph.

As the green arrows show, the modes at the extremes of the graph tend towards a high level of efficiency and productivity, the best trend (green arrows). The "AUTOMOBILE 2014/1930" modal, on the other hand, shows a strong tendency towards the low level, the worst tendency (red arrow).

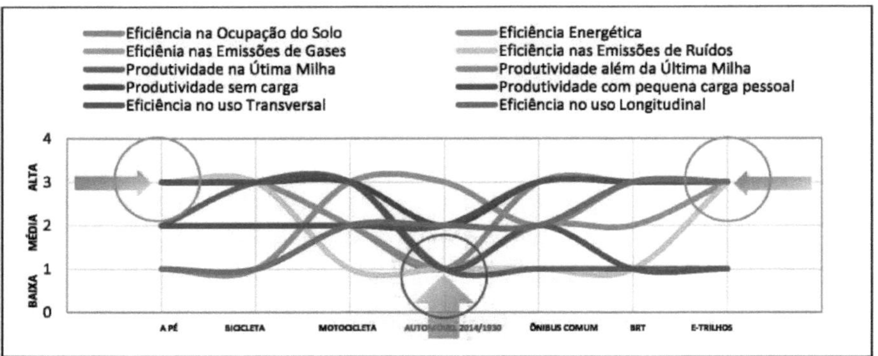


Figure 4: Urban mobility modes for people - scatter diagram. Source: The author (2014), taking into account the references in section 3.2.1.

In order to visualise exactly how many occurrences there are at each level for each mode, figure 5 was drawn up based on the data in figure 4. It was considered that from five occurrences onwards the trend is extremely marked, which proves the comments made in figure 4.

| Modals | | | | | | | E-Rails |
|---|---|---|---|---|---|---|---|
| Occurrences | On foot | Bicycle | Motorbike | Car 2014/1930 | Regular Bus | BRT | |
| HIGH | 6 | 7 | 4 | 1 | 3 | 5 | 8 |
| MEDIA | 2 | 1 | 4 | 3 | 4 | 1 | 0 |
| LOW | 2 | 2 | 2 | 6 | 3 | 4 | 2 |

Figure 5: Number of incidents at each level by mode of urban mobility. Source: The author (2014), considering the dispersion data in figure 4.

As an immediate conclusion from analysing figure 4, it can be said that in order to choose the best modes available today for urban mobility, one should tend towards the extremes, as shown in the scatter plot in figure 4, and/or improve the indices of the modes that are most suitable for urban mobility.

are scoring at low levels so that they also have higher levels of efficiency and productivity, since all modes are indispensable and affect sustainability and quality of life in urban traffic.

Graphs will then be presented with each parameter individually so that each can be visualised in isolation in relation to all modes.

## 3.3 Cars and land use

It can be seen that the "2014/1930 car" mode is the mode with the worst index in figure 6 in relation to the land use efficiency parameter.

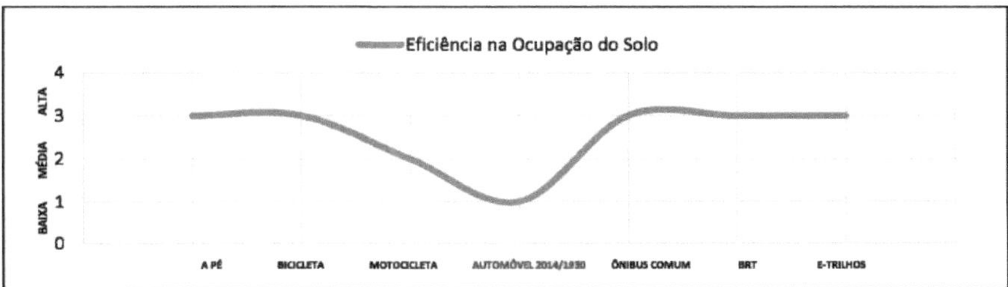


Figure 6: Urban mobility modes for people - Land use efficiency. Source: The author (2014), considering the references in section 3.2.1.

YAMAWAKI (2014) explains in her article "Smaller cars to help relieve traffic", from the point of view of an urban architect, that the number and size of cars in relation to the size of the roads and spaces available, and the rate of practically individual use of each vehicle, means that there is a marked disproportion and waste in this mode of transport as it stands. "The model is a legacy of the Roman chariot period, which influenced animal-drawn carts. This size has moulded our current spaces with large areas for vehicular flows and smaller stretches for pedestrians."

SIGEL (2013) in his article "How to empty the streets of cities" concludes that just as it happens in the design of an architectural project - the search for the best efficiency and comfort of a space - with the compatibility of various disciplines, the time has come to give the deserved importance to the space intended for vehicles, such as the efficient use of technology, both at the time of design and when operating the space.

The car in its current proportions in cities, and according to the levels of use already explained, increases the inefficiency of the modal and hides extremely compromising figures in terms of efficiency in land occupation that are not taken into account, such as the fact that for every new car sold on the market, it is necessary to provide a new free space in the city of around 50 m² to absorb this new vehicle and allow it to integrate into traffic. This 50 m² means that this new vehicle must have around 17 m² for its exclusive private garage, 17 m² in road space and 17 m² in public or private car parking spaces in public or private commercial or service premises. In this way, they clog up cities with useless demands, because they are not proportional to the space people need to get around.

In order to get a clear picture of the efficiency of land occupation of the "AUTOMOBILE 2014/1930" modal, the IPPUC website presents some very interesting statistical data. Taking an example from the traffic count survey data and selecting and freezing as a photo a city street and analysing the efficiency of land occupation in the survey carried out in Curitiba/PR on 09/08/2012 at the intersection of Brigadeiro Franco Street and Alameda Doutor Carlos de Carvalho, from 5pm to 8pm, it can be seen that 5558 cars, 15 buses and 11 trucks passed through this intersection on Brigadeiro Franco Street. Considering that the street has 3 lanes, each almost 3 metres wide, that each bus is 15 metres long and each popular car is 4 metres long, it appears that 75 metres of street would be needed to accommodate all the buses and 7500 metres of street to accommodate all the cars. Considering that buses travel with a capacity of 110 passengers and cars with an average of 1.5 people (according to the statistics that vary from 1.46 to 1.59 in Technical Bulletin 31p2 - Vehicle Occupancy Survey - on the website www.cetsp.com.br), using a land use efficiency ratio, we can deduce that bus transport houses 2.44 people/m$^2$ and cars house 0.12 people/m$^2$ . The figures are compatible with the individual analysis of each vehicle in each mode of transport, which would have the following land occupation rates per person, demonstrating the poor rate of the "2014/1930 car" mode of transport, as shown in table 1.

Table 1: land use index for each means of transport.

| Means of transport | Length (m) | Width (m) | Number of people per vehicle/modal | land occupation/person index |
|---|---|---|---|---|
| on foot (walking) | 0,5 | 0,5 | 1 | 0.25 m /person$^2$ |
| bicycle | 1,5 | 0,8 | 1 | 1.2 m /person$^2$ |
| Motorbike | 2,0 | 0,8 | 1 | 1.6 m /person$^2$ |
| Car_2014/1930 | 4,0 | 2,0 | 1,5 | 5.34 m /person$^2$ |
| Buses | 15,0 | 2,5 | 110 | 0.34 m /person$^2$ |
| BRT (bi-articulated) | 25,0 | 2,6 | 230 | 0.28 m /person$^2$ |
| Metro (6 carriages) | 130,0 | 3,15 | 2000 | 0.21 m /person$^2$ |

Source: The author (2014), considering the references for each item.

Figure 7, which shows the comparative relationship between the number of passengers on a bi-articulated bus and the number of "2014/1930 cars" needed to house the same number of passengers, clearly shows the inefficiency of the land use of the "2014/1930 car", which in practice is even worse, because in the example above it was considered that each car is carrying 4 passengers, which in practice is not true, because as already reported, in reality each car travels with 1.5 people on average, which would give a total of approximately 153 "2014/1930 cars" with inappropriate use/dimensions.

Figure 7: Passenger capacity ratio between buses and cars SOURCE: URBS-2009 ELABORATION: Monitoring Sector - 2010

It can be concluded that in the "AUTOMOBILE 2014/1930" mode there is extreme inefficiency in land use for mobility due to the size of the vehicle and the almost individual use of the vehicle. And the problem is even more acute when the vehicle is used by a single person, and even if the vehicle is larger than the dimensions of the popular standard, such as the SLIVs widely used today, which can reach a land use index of 11 m$^2$ /person.

## 1.4  0 cars and energy efficiency

Considering the data obtained in the interview in the appendix that the energy efficiency of cars can vary between 7 and 13 per cent, and according to the information from INEE in figures 9 and 10 that "from well to wheel" the vehicle uses 10 per cent of the energy for mobility and 12 per cent "from pump to wheel", and considering that cars are used to move an average of 1.5 people, and with the land occupation rate obtained in table 1, the energy efficiency of cars is at a low level as shown in figure 8.

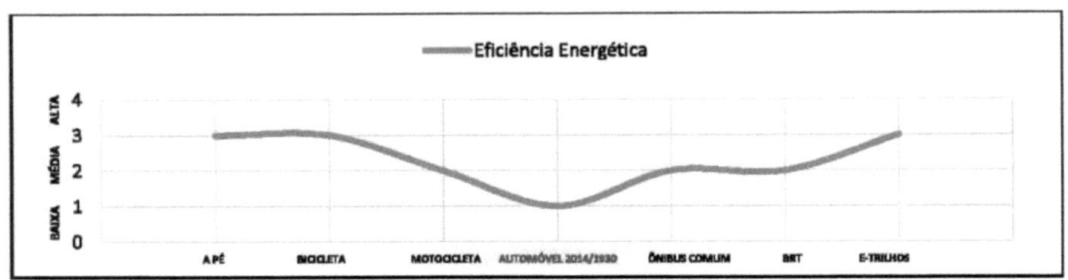


Figure 8: Urban mobility modes for people - energy efficiency Source: The author (2014), considering the references in section 3.2.1.

Figures 9 and 10 also show that something that also greatly compromises energy efficiency, among other things, is inertia, the ***"stop and go"*** of congested traffic, which is responsible for 56 per cent of the vehicle's energy loss under the "pump to wheel" analysis. This performance is confirmed by IPPUC statistics measuring the speed and travel time of vehicles on the Curitiba Ring Road, which shows an average speed of around 20 km/h in urban traffic, which proves that the vehicle stops and goes several times in traffic.

Figure 9: Energy efficiency of the wheel well.
Source: INEE - National Institute for Energy Efficiency

Figure 10: Energy efficiency of the wheel pump.
Source: INEE - National Institute for Energy Efficiency

After more than a century of engineering, experts are embarrassed by the inefficiency of vehicles on the road. At least 80 per cent of the fuel energy they consume is lost, mainly in heat and the exhaust of engine gases, so that less than 20 per cent is currently used for wheel traction. According to Hawken (1999), 95 per cent of the resulting force is used to get the car moving, with only 5 per cent moving the driver. This means that only 1% of the fuel's energy is used to satisfy modern man's need for locomotion.

## 3.5 The car and the efficiency of gas and noise emissions

In figure 11, in terms of both gas emissions and noise emissions, the modes have the same efficiency levels for both parameters, especially for the internal combustion engine system and equally with the non-motorised and electric ones.

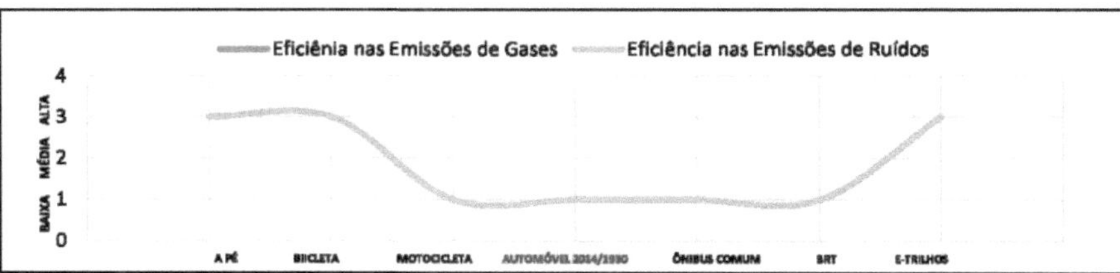


Figure 11: Urban mobility modes for people - emissions efficiency. Source: The author (2014), considering the references in section 3.2.1.

According to ANTP's 2011 general report, pollutant emissions are estimated considering two types: local pollutants (Carbon Monoxide (CO), Hydrocarbons (HC), Nitrogen Oxides ($NO_X$ ), Sulphur Oxides ($SO_X$ ) and Particulate Matter (PM)) and greenhouse pollutants (Carbon Dioxide (CO2)). Total emissions represent the sum of emissions of these two types.

Table 2 of ANTP's 2011 general report shows that the bus, motorbike and car modes generate gas emissions, and cars are responsible for 61 per cent of the amount emitted. Considering the analyses above regarding the inefficiency of cars, especially due to their underutilisation because of their size and energy inefficiency, this emissions factor increases the inefficiency effect for this vehicle.

Table 2: Total vehicle emissions (million tonnes/year) - 2011

| System | Million tonnes | Share (%) |
| --- | --- | --- |
| municipal bus | 7,3 | 24 |
| metropolitan bus | 2,8 | 10 |
| *Public Transport - Total* | *10,1* | *34* |
| Auto | 17,6 | 61 |

| | | |
|---|---|---|
| **Motorbike** | **1,4** | **5** |
| *Individual Transport - Total* | *19,1* | *66* |
| **Total** | **29,2** | **100** |

Source: table 40 of ANTP's 2011 general report.

CONAMA Resolution 272 of 14 September 2000, considering that vehicles with combustion engines generate noise emissions, and also considering the need to reduce noise pollution in urban centres, determined that vehicles must reduce noise emissions, which on the scale of the modes in question classifies modes with combustion engines at the low efficiency level in relation to non-motorised ones and electric ones that reach maximum efficiency levels.

### 3.6  The car in relation to the last mile

Figure 12 shows that non-motorised modes and the motorbike, which is also an agile vehicle in traffic, have high productivity in the last mile, which refers to the activities to be carried out over the shortest distances and the most tangled routes. On the other hand, motorised modes, including motorbikes, have better productivity when it comes to travelling beyond the last mile, both because of the longer distances and the lower levels of traffic volume. But the main thing to note is that the 2014/1930 car has a good level of productivity beyond the last mile and low productivity in the last mile, which is justified by all the previous considerations, especially its size and the low utilisation rate of its capacity resources.

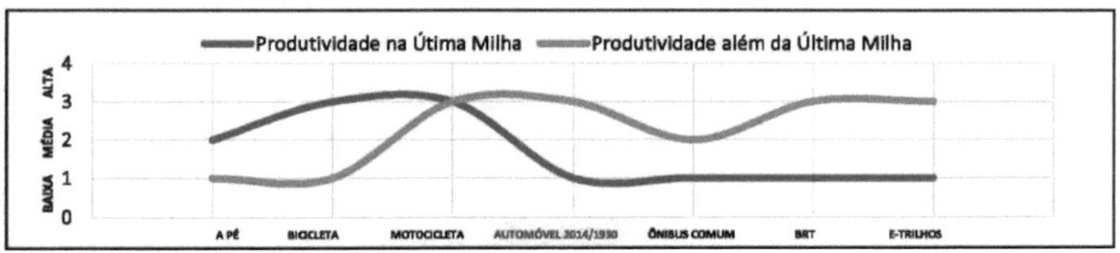


Figure 12: Urban mobility modes for people - productivity in distances. Source: The author (2014), taking into account the references in section 3.2.1.

### 3.7  The car in relation to transverse and longitudinal displacement

Considering that the transverse journey takes place within the last mile and the longitudinal journey is the one that starts within the last mile and travels the stretches beyond the last mile, figure 13 again shows curves similar to figure 12 in relation to the modes.

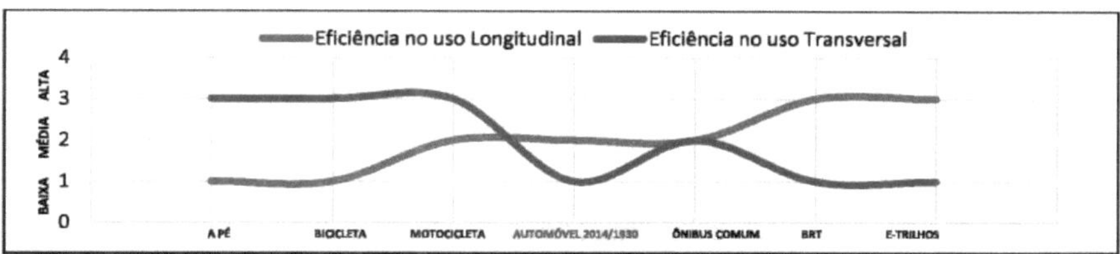


Figure 13: Urban mobility modes for people - efficiency in the direction of use.
Source: The author (2014), considering the references in item 3.2.1.

Transverse journeys are made within the last mile in the traffic tangle of the urban nerve centre(s). Practical examples are the common journeys between the home or business and the market, and the shopping centre, the children's school, the gym, the restaurant.

## 3.8  The car in relation to Productivity without load and with personal load

When the Productivity of the modes is analysed in relation to travel without a load or with a small personal load, as shown in figure 14, it can be seen that these parameters can only have a negative influence on non-motorised and individual mobility modes, and have no influence on collective modes. The reason for this is that non-motorised modes suffer from loads due to their volume and weight, which can make travelling difficult, and individual mobility modes suffer from the fact that handling loads usually depends on the driver and requires the use of car parks, which can be difficult in the last mile. In the case of public transport, this factor has no influence due to the independence between the driver and the person carrying the load.

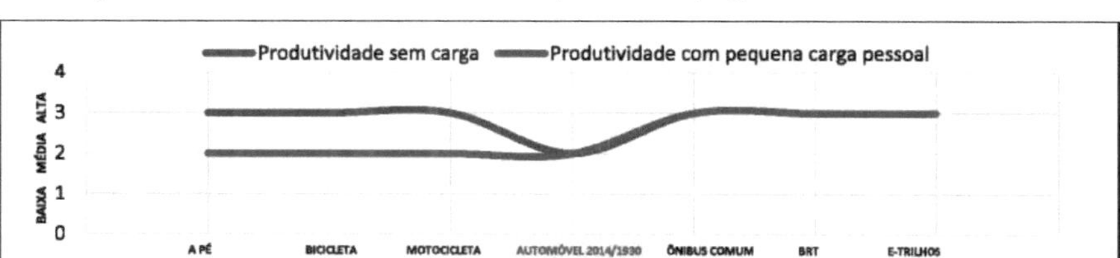


Figure 14: Urban mobility modes for people - productivity in relation to freight. Source: The author (2014), taking into account the references in section 3.2.1.

# CHAPTER 4

## WHY A 1930 MODEL CAR?

Once you know what the car is, its characteristics and how it is positioned in the environment and in urban mobility, you can compare, with a critical eye and in the evolution of time, the standard of the car in relation to the necessary or desired standards to meet the needs of individual urban mobility demands, inversely to what was seen in the previous chapters, in order to harmonise the car with sustainability for the city and the planet and quality of life in the city.

Figure 15, ***"Demand for URBAN TRAVEL x Development of URBAN CARS",*** was created based on two references: **YEAR** and **YEAR of "Model/Standard/Style".** The **Urban Commuting Demand** line has varied over time in relation to the "Model/Standard/Style" in line with the growth of the urban population, which, the greater it is, the greater the demand for commuting, the greater the need for people to move around. This curve has been accentuated since 1920 and more strongly since the 1960s and 1970s with the strong rural exodus and the strong growth of the urban population, as already pointed out in the introduction to the 2010 IBGE Demographic Census, which reports that in 1920 10% of the population lived in urban areas and in 1970, 55.9% and in 2010, 81.22%. It can also be seen that at some points on the line the x and y ratio in figure 15 shows the same data, while at other points it shows that the value of y has not kept pace with the value of x, i.e. that the time-year has progressed punctually, but the model-year has lagged behind, i.e. the model has not evolved proportionally in both references over time, probably because the physical spaces available in the cities and their surroundings allowed gaps in evolution until around the 1960s or 1970s, when from then on the x and y references kept evolving together in time and in the pattern in relation to the demand for travel, making this curve ascending.

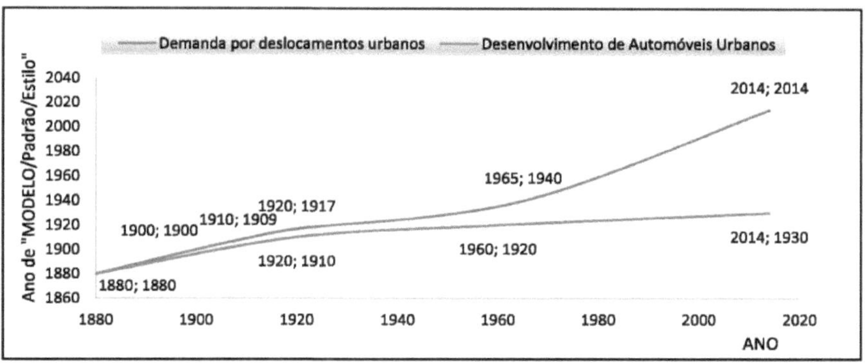


Figure 15: Demand for urban travel x Development of urban cars. Source: The author (2014), simply an illustrative trend graph without precise dates.

The line of ***development of URBAN AUTOMOBILES***, on the other hand, has varied over time in terms of "Model/Pattern/Style" according to the evolution of the automobile. It was invented at the end of the 19th century, underwent structural changes practically until the 1930s and since then has hardly ever changed its standard in terms of dimensions and base, evolving only in terms of material technologies, aerodynamics and on-board technologies (features and safety) as can be seen in figure 16.

Figure 16 shows the evolution of on-board controls in cars, with the increase in complexity over the years. It shows that from the 1900s to the 1960s, vehicles were completely mechanical. From the 1960s to the 1980s, on-board electronics were introduced. From the 1980s onwards, on-board controls were adapted.

navigation and stability control electronics, remote sensing, communications and driving interventions. But what is clearly noticeable, just by looking at the images in the figure, is that when the car acquired its "shape" in terms of pattern and dimensions around the 1930s, it never changed that pattern and basic characteristics, assuming that today it is the 2014 model year 1930 car from the point of view of urban needs.

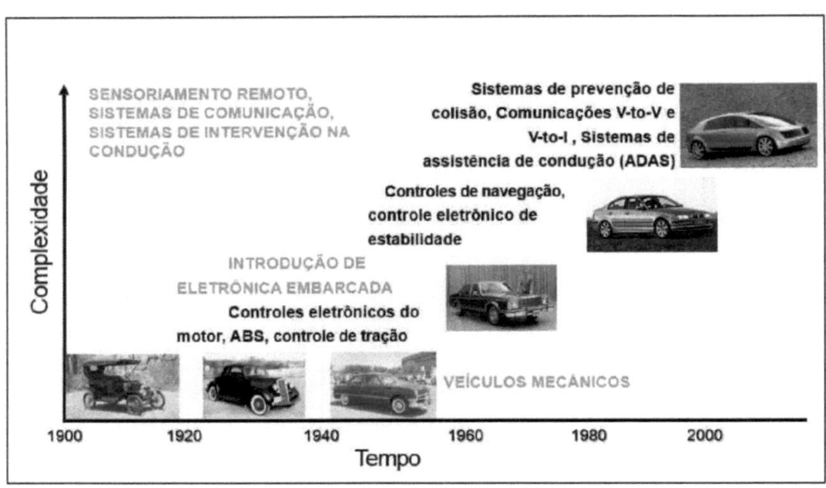


Figure 16: Evolution of car controls.

It is understood that the line of evolution of the car has hardly been accentuated in relation to its "Model/Standard/Style" due to the fact that it is being analysed from the point of view of adapting to the urban environment, which from the 1930s and 1970s onwards began to more strongly demand vehicles for urban mobility that contributed to sustainability and quality of life, i.e. efficient in relation to land use, energy, emissions, agility in traffic, productivity and in terms of available capacity and dimensions versus rate of use. And the car has not kept up with the needs of urban environments in these respects, which is represented in the graph by the lag between the values of x and y in the variable, ultimately showing the car as being manufactured in 2014, but modelled in 1930!

It can be seen in figure 15, with the detachment or distance between the two trend lines, the *demand for urban mobility* and the *development of* suitable urban *vehicles*, that in 2014 there was a high demand for urban commuting, but with 1930s-standard vehicles, i.e. out of step with the demands of 2014. They are therefore unsuitable for use in urban environments.

Ideally, the "trend behaviour" of the graph in figure 15 would be similar to that of the graph in figure 17, with the trend lines always relatively "glued" or close together, as can be seen.

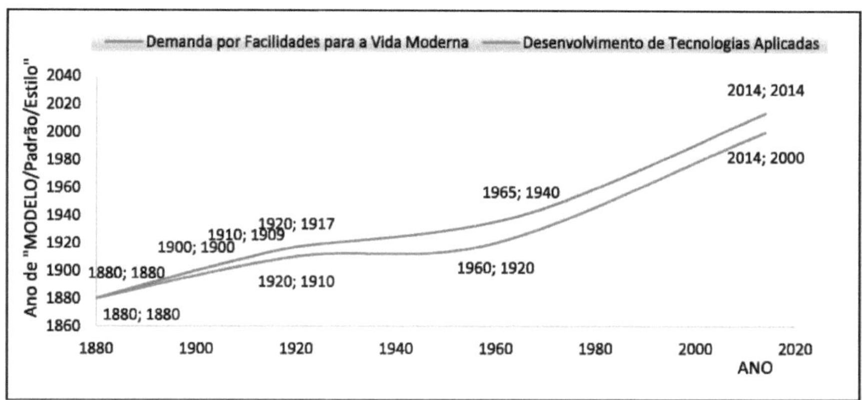


Figure 17: Demand for facilities for modern life x Development of applied technologies. Source: The author (2014), simply an illustrative trend graph, without precise dates.

The graph in figure 17 shows that the trend lines of the variables have always remained relatively close, with some lag in the development of applied technologies in relation to the demand for facilities for modern life, which is normal, as for the most part the market has always made solutions available after the demand has occurred. It should be noted again that from the 1960s onwards, with the high rate of urbanisation and greater generation of wealth, and with the advent of electronic and computer resources, the curves rose sharply, but the important thing is that they have always evolved in parallel and relatively "glued together", thus ensuring that the needs of one are met by the resources of the other.

Quality of life has generated demand for technologies applied to products that bring "facilities" to people's lives in cities.

Technology has made available products with added value and suited to modernity, such as smartphones, mobile phones, ultrabooks, tablets, audio, video and data communications, lifts, escalators, air conditioners, wearables today and the use of nanotechnology, and the infinity of solutions in products and materials for people's productivity and well-being.

In the area of technologies applied to modern living facilities, trends have evolved in tandem, i.e. as quality of life has required innovation, the benefits have come in the form of new products and equipment that are up to date with demand. Technology is doing its bit, such as the internet and audio and videoconferencing systems that shorten distances, widen roads and reduce or even eliminate journeys. But the development of cars, especially since the 1970s and in relation to the real needs of the urban centres whose factors have been analysed, has almost entirely been restricted to on-board technologies and more as an advance in comfort and safety items than as solutions to the demands of the complex urban mobility environment, thus distancing the car from the sustainable urban profile and quality of life.

On the basis of the arguments and facts set out above, it can be seen that, in general terms, the car industry has managed to deliver good cars to the market over the last 100 years or so for extra-urban and weekend commuting, but that today they are not suitable for productive and efficient individual urban mobility!

# CHAPTER 5

CONCLUSION

The conclusions will consider the arguments and facts presented and even propose future studies and projects with the aim of deepening, broadening and advancing the understanding and solutions of urban mobility and its components.

## 5.1  Author's analysis

The urban planet, as mentioned in the introduction, requires commitment and action from government bodies, industry and society if we are to have a sustainable planet and a city with quality of life. The individual urban mobility vehicle has a fundamental, complementary and essential role to play in satisfying people's comprehensive movement needs in cities, but it needs to evolve in order to harmonise with the city.

The car cannot simply be excluded from the context of the modes available for urban mobility, otherwise there will be a lack of a complementary and necessary resource and alternative for people's cross-cutting journeys in the urban environment. Particularly because users say they like their cars and, more than that, they want to have means of transport available that provide them with speed, comfort and independence in their journeys, as well as the freedom to be able to go to different places on the same journey. Urban mobility can be compared to a Swiss Army knife in which the various modes are all indispensable and irreplaceable for realising the diversity of people's movement needs in the city.

What has been identified as the problem with this mode of transport is that the car industry is providing the market with the "2014 model year 1930 car" when it should be providing a 2014 model year car that is compatible with the needs of the urban planet.

The car, in its own sense of the word and not in the standard established in the 1930s, is an essential vehicle in individual urban mobility for transversal journeys in urban centres, because it is irreplaceable as a mode of transport integrated into the network of modes of transport for journeys such as going to your favourite bakery, your favourite market, your favourite cinema or picking up the kids from school.

The automobile industry is proud of the advances made in cars, as seen in figure 15, but it is understood that they are not enough in terms of the needs of urban mobility, the sustainability of the planet and quality of life. It can be concluded that the car's value proposition from the perspective of today's urban reality has lost its justification and leaves a problem of loss of identity for its use within the scope defined in this analysis.

Quoting DIAS (2003), "...it's curious that transport planners have recognised that building more roads doesn't necessarily solve traffic problems (it's like

buy wider trousers to solve the problem of obesity), but continue to cut down green areas to make way for more and more roads."

"Comfort, speed and independence are some of the indisputable advantages of cars. However, the indirect costs of this type of dependency are becoming apparent: road maintenance and construction, congestion that undermines economic productivity, high levels of energy consumption with their respective environmental and economic impacts, increased atmospheric pollution" (DIAS, 2003).

And as a transformative message, we can look to a sentence in Curitiba's first masterplan that says: "At the heart of Curitiba's urban transformation is the concept that the human being is the measure of all things and the city must be conceived as a meeting place for citizens". The car needs to consider the human being as its measure.

5.2  Perspectives and suggestions for future work

It is suggested that dedicated studies and research be carried out, including in the field, in order to deepen the analysis of indicators on the limits of car behaviour in the last urban mile, with and without personal cargo, in transversal and longitudinal journeys, even contrasting with data obtained in the laboratory by manufacturers.

Studies are also suggested to survey historical data and proposals for cultural evolution based on the concepts of sustainability in the urban mobility environment.

It is also suggested that studies be carried out to achieve the ideal car with on-board technologies for new urban cars whose year of manufacture and model year are always the same as the current calendar year, at the same time as traffic management evolves.

As a follow-up to this study, the aim is to develop a project for the city of Curitiba, with the effective and multidisciplinary participation of academia, government and industry, with the implementation of a *virtual model with 4D immersion,* presenting proposed advances for the urban car and the respective integration and evolution of traffic management.

# CHAPTER 6

REFERENCES


NATIONAL ASSOCIATION OF PUBLIC TRANSPORT -ANTP. Information System of the Urban Mobility - General Report 2011-December/2012. Available at: <http://www.antp.org.br/website/produtos/sistema-de-informacaoes-da-mobility/show.asp?ppgCode=63451652-6DEE-4CCE-81D5-1162F86ClC19> Accessed on: 21 Mar. 2014.

BEGGS, Clive. Energy: management, supply and conservation. Woburn MA: Butterworth - Heinmann, 2002.

BRAZIL. Congress. Law no. 10.257, of 10 July 2001, City Statute.

BRAZIL. Congress. Law no. 12.587, of 3 January 2012, National Urban Mobility Policy.

CLAUDINO, João B. Commercial diesel engines in Brazil and the environment. Curitiba, Núcleo, 1999.

SIBRT BEST PRACTICE CONGRESS IN LATIN AMERICA, 3, 2013, Belo Horizonte, MG. The car is the cigarette of the future. Available at: <http://congresosibrt.org/pt/noticias/159/o-carro-e-o-cigarro-do-futuro> Accessed on: 12 Feb. 2014.

CONPET in Transport. In: CONPET Programa Nacional da Racionalização do uso dos derivados do petróleo e do gás natural. Available at: <http://www.conpet.gov.br/portal/conpet/pt_br/pagina-inicial.shtml> Accessed on: 21 Nov. 2013.

DEBATE: Cars and Consumption - Mobility, Emissions and Energy Efficiency. IDEC Brazilian Institute for Consumer Defence. Available at: <https://www.idec.org.br/mobilize-se/evento/debate-automovel-e-consumo> Accessed on: 17 Feb. 2014.

DENATRAN National Traffic Department. VEHICLE FLEET - Fleet Yearbook 2014. Available at: <http://www.denatran.gov.br/frota.htm> Accessed on 10 Feb. 2014.

DIAS, R. Americans are driving more and more. Folha de São Paulo, São Paulo, 11 May. 2003. Money. Notebook B.

DUARTE, Fábio; LIBARDI, Rafaela; SÁNCHEZ, Karina. Introduction to urban mobility. Curitiba: Juruá, 2009.

Traffic jams in São Paulo caused losses of R$40 billion in 2012. Bom Dia Brasil. Available at: <http://g1.globo.com/bom-dia-brasil/noticia/2014/Q3/congestionamentos-de-sp-causaram-

prejuizo-de-r-40-bilhoes-em-2012.html> Accessed on: 12 Mar. 2014.

**Transport Statistics.** Available at: <http://www.urbs.curitiba.pr.gov.br/transporte/estatisticas>
Accessed on: 05 Feb. 2014.

HAWKEN, Paul... [et o/.]. **Natural Capitalism: Creating the Next Industrial Revolution.** New
York: Little Brown and Company, 1999.

**NATIONAL INSTITUTE OF METROLOGY, QUALITY AND TECHNOLOGY INMETRO.** Ordinance
no.?
549, of 25 October 2012. Available at: <
http://www.inmetro.gov.br/legislacao/rtac/pdf/RTAC001922.pdf> Accessed on: 21 Nov. 2013.

JALURIA, Yogesh. **Design and optimisation of thermal systems.** Singapore: McGraw-Hill, 1998.

LE CORBUSIER. **Urban Planning.** São Paulo: Perspectiva, 2013.

LEITE, Carlos. AWAD, Juliana di Cesare Marques. **Sustainable Cities, Smart Cities - Sustainable
Development on an Urban Planet.** São Paulo: Bookman, 2012.

LERNER, Jaime. **Urban Acupuncture.** Rio de Janeiro: Record, 2003.

MACHADO, Anaxágora Alves. Noise pollution as an environmental crime. **Jus Navigandi,**
Teresina, year 9 (/revista/edicoes/2004), n. 327 (/revista/edicoes/2004/5/30), 30
(/revista/edicoes/2004/5/30) May (/revista/edicoes/2004/5) 2004 (/revista/edicoes/2004).
Available at: <http://jus.com.br/artigos/5261> Accessed on: 17 Dec. 2013.

**The Future of the Cities of the Present.** REVISTA CRN BRASIL, São Paulo, It Midia, Issue 363, p.
31-37, July 2013.

PANESI, André R. Quinteros. **Fundamentals of Energy Efficiency.** São Paulo, Ensino Profissional,
2006.

**CNI-IBOPE survey: portraits of Brazilian society: urban commuting (August 2011).**
Available at <www.cni.org.br> Accessed on: 05 Mar 2014.

**Traffic Count Survey.** In: IPPUC - INSTITUTE FOR RESEARCH AND PLANNING
CURITIBA URBAN CENTRE. Available at: <http://www.ippuc.org.br/default.php> Accessed on: 10
March 2014.

**Survey Measuring vehicle speed and journey time 2012/2013 - Ring Road.**
In: IPPUC - INSTITUTO DE PESQUISA E PLANEJAMENTO URBANO DE CURITIBA. Available at:
<http://www.ippuc.org.br/default.php> Accessed on: 10 Mar. 2014.

**São Paulo Metropolitan Region Mobility Survey 2012 - Main results household survey
December 2013.** Available at:

<http://www.metro.sp.gov.br> **Accessed on:** 17 Mar. 2014.

**Vehicle Occupant Survey.** In: CET - São Paulo Traffic Engineering Company. Technical Bulletin no. 31p2. Available at: <http://www.cetsp.com.br> Accessed on: 17 Mar. 2014.

PINTO JÚNIOR, Helder Queiroz... [et **al]. Energy Economics: economic foundations, historical evolution and industrial organisation.** Rio de Janeiro: Elsevier, 2007.

**PNEF National Energy Efficiency Plan.** In: MINISTRY OF MINES AND ENERGY.
Available at:
<http://www.mme.gov.br/mme/galerias/arquivos/noticias/2011/Plano_Nacional_de_Eficixn cia_Energxtica_-_PNEf_-_final.pdf> Accessed on: 21 Nov. 2013.

**Curitiba's first masterplan dates back to the 1960s.** Curitiba City Hall News Agency - published on 14/03/2014 11:21:00. Available at:
<http://www.curitiba.pr.gov.br/noticias/primeiro-plano-diretor-de-curitiba-e-da-decada-de-1960/32276> Accessed on: 20 Mar. 2014.

**Brazilian Vehicle Labelling Programme.** In: INSTITUTO NACIONAL DE METROLOGIA, QUALIDADE E TECNOLOGIA INMETRO. Available at: <http://www2.inmetro.gov.br/pbe/> Accessed on: 20 Nov. 2013.

**CONAMA RESOLUTION no[5] 272,** of 14 September 2000 Published in DOU no 7, of 10 January 2001, Section 1, page 24. Available at:
<http://www.mma.gov.br/port/conama/legislacao/CONAMA RES CONS 2000 272.pdf>
Accessed on: 24 Mar 2014.

ROGERS, Richard. **Cities for a small planet.** São Paulo: Gustavo Gili, 2012.

SALDIVA, Paulo... [et *al.]. ***Environment and Health - the Challenge of the Metropolises.* São
Paulo
Paulo: Health and Sustainability Institute, 2010.

SIGEL, Ricardo. How to empty city streets. **Gazeta do Povo,** Curitiba, 15 December 2013. p. 01.

**VEHICLES.** COMIL Ônibus S/A. Available at: <http://www.comilonibus.com.br/site/veiculos>
Accessed on: 15 Apr. 2014.

**INEE's vision of energy efficiency.** In: NATIONAL INSTITUTE FOR ENERGY EFFICIENCY - INEE.
17/8/2007; Pietro Erber, Jayme Hollanda, Osorio de Brito and Orlando Puppin. Speech given at the
"Joint Meeting of INEE's Board of Directors, Advisory Board and Executive Board". Available at:
<http://www.inee.org.br/down loads/sobre/Conselho 17Ago final.ppt> Accessed on: 12 Apr. 2014.

YAMAWAKI, Yumi; Tomaz, Juliana. Smaller cars to help ease traffic. **Gazeta do Povo,** Curitiba, 11

May 2014. Real Estate, p. 01.

# GLOSSARY

- **On foot =** referring to pedestrians walking to get around;

- **AUTOMOBILE =** 4-wheel motorised vehicle for up to 8 people + driver;

- **BRT =** Bus Rapid Transit;

- **Nervous Urban Centre =** central agglomeration(s) in the city;

- **E-Rail =** means of transport on rails, with an electric motor, such as trains and metros;

- Efficiency = efficiency or yield refers to the relationship between the results obtained and the resources employed;

- **Energy Efficiency =** the best ratio between the amount of final energy used and the amount of goods produced or services rendered;

- **Efficiency in Gas Emissions =** is the lowest generation of waste gases for a good produced or service carried out;

- **Noise Emission Efficiency =** is the lowest noise generation for a good produced or service carried out;

- **Land Use Efficiency =** the best ratio between the amount of land used by a vehicle or means of transport and the volume it transports;

- **Efficiency in longitudinal use =** the greater agility and results obtained by the means of locomotion starting from the "central zero point" and directly reaching its objective far from the central area by travelling along an imaginary straight line;

- **Efficiency in transversal use =** the greater agility and results obtained by the means of locomotion within the nerve centre and reaching one or more points in the same central area by travelling along one or more imaginary irregular circular lines;

- **Noise Emissions =** noise emitted by means of locomotion;

- **Gas emissions =** gases emitted by means of locomotion;

- **Energy =** what enables work to be done;

- **Extra Urban =** area outside the city limits or the urban nerve centre;

- **Longitudinal Mobility =** movement made by the means of locomotion starting from the "central zero point" and directly reaching its objective far from the central area by travelling along an imaginary straight line;

- **Transversal Mobility** = movement carried out by the means of locomotion within the nerve centre and reaching one or more points in the same central area along one or more imaginary irregular circular lines;

- **Urban Mobility** = travelling within the city area;

- **Urban Freight Mobility** = moving freight within the city area;

- **Collective Urban Mobility** = means of transport to move large volumes of people within a city area;

- **Individual Urban Mobility** = means of transport to move between one and nine people within the city area;

- **Motorised Urban Mobility** = travel within the city area by motorised transport;

- **Non-motorised Urban Mobility** = travel within the city area by non-motorised transport;

- **Urban Mobility of People = the** movement of people within the city area;

- **Urban Mobility Modes** = means of transporting people or freight within the city area;

- **Land Use** = the physical space used by the means of transport, normally relative to its circumference when parked or in motion;

- **Productivity** = is the best relationship between production and the factors of production used; it is the direct relationship between what is produced (time) and what should be produced (time) its result is given as a percentage (%);

- **Productivity beyond the last mile** = the best ratio between production and the factors of production used when travelling outside a nerve centre area;

- **Productivity with low personnel load** = the best ratio between production and the factors of production used obtained when travelling with low personnel load;

- **Unladen productivity** = the best ratio between production and the factors of production used obtained when travelling unladen;

- **Last mile productivity** = the best ratio between production and the factors of production used, obtained when travelling within a nerve centre area;

- Quality of Life = is the measure of a human being's living conditions, involving physical, mental, psychological and emotional well-being, social relationships such as family and friends, as well as health, education and other life circumstances. In relation to urban mobility, it's the facilities and comfort that modes of transport provide to meet your travelling needs and with the lowest level of consequences that are harmful to your health, well-being and enjoyment;

- Sustainability = the ability to carry out mobility while guaranteeing and providing quality of life for people and minimising the harmful effects on the planet;

- Last Mile = the central space of the traditional city with the highest volume of traffic of people and vehicles, extending to the limits of this volume with an imaginary circle. In modern cities, we can see the creation of various nerve centres, connected or not, considered from the centre of these "poles" to be various "last miles";

- Urban = relating to the city;

- Longitudinal use = moving from the "central zero point" and reaching a point directly away from the central area along an imaginary straight line;

- Transversal use = locomotion within the nerve centre and reaching one or more points in the same central area by travelling along one or more imaginary irregular circular lines;

APPENDIX - Interview on Vehicle Energy Efficiency

Interview conducted on 18 February 2014 with Luiz Osório Vendrami Trentini, a mechanical engineer and civil engineer with a postgraduate degree in physics from the Federal University of Paraná and extensive experience in the automotive sector, having worked for Volvo as industrial quality manager, Renault as quality manager for MERCOSUR, Nissan as manager of Cost Control and Investment, and Denso as senior production process engineer. He is also one of the inventors of the *TRIKKE, an* inclinable tricycle that moves on the principle of angular momentum with more than 600,000 units sold worldwide and chosen by Time magazine as one of the best inventions in the USA in 2002.

The aim of the interview was to investigate the views of a highly experienced engineering professional on the <u>ENERGY EFFICIENCY</u> of cars currently used for <u>INDIVIDUAL URBAN MOBILITY</u> in Brazil, in day-to-day use during commercial traffic periods.

Iª question)

Is today's car industry developing vehicles for mixed use (urban and interurban, and in some cases on and off road)? At what cost does it achieve this elasticity? What characteristics does this imply? And are there gains and/or losses in this implementation pattern?

Answer: Yes, the car industry develops vehicles for mixed use. There are impacts when the car is designed for multiple use (city and road) in the aspects of engine power, vehicle size and weight, and energy efficiency. The cost is impacted upwards, as the vehicle is designed for higher speeds, more powerful engines, more elaborate suspensions, more aerodynamic bodies and therefore larger, heavier bodies, etc. Environmentally, there are losses, as 95% of the time vehicles are used is in urban areas, where the average speed is around 30 km/h in small towns and 20 km/h in large cities, and even less in megalopolises. The maximum speed in cities is 60 km/h, so the streets are full of large cars carrying one person, the engines are only working at 8% of their maximum power, which represents an energy efficiency of around 10% in relation to the energy actually needed to give the same mobility to a 70 kg person on foot.

2ª question)

To analyse Energy Efficiency issues, would it be appropriate to take a representative sample of the

best-selling "popular compact" car in Brazil and an example of an SUV that is widely sold in Brazil? Does this represent a sample of the largest volume of vehicles used in urban centres?
Answer: Yes

3ª question)
Then select an example of each style and present the Energy Efficiency data obtained in the practice applied according to the conditions of use defined above. Answer:

Analysing the efficiency of human mobility in cities

Mobility is one of the inalienable rights of living beings. Man is guaranteed the right to come and go by constitutional right. The value of the freedom to choose which route a person will take is incalculable. Because of these factors, the car has become the alternative for mobility in modern society, and urban planning and city streets are guided by the king of the city, "Mr Car", because with it people can, up to certain limits, decide where to go, which route to choose, how fast to travel, etc. According to research carried out by Andreas Schafer et al,[1] it was found that people spend a global average of 1 hour and 15 minutes travelling every day, and what varies is the mileage travelled per person. As the distance travelled is speed multiplied by time, and it is known that speed is a function of the technology used in the transport vehicle and the infrastructure, it is clear that the distance travelled increases proportionally with the GDP per capita of the economic region where the individual lives. For example, in the USA the average is 20,000 Km/year x capta or 55 Km/day x capta. In Brazil it's 8,000 Km/year x capta or 22 Km/day x capta and in China it's 3,000 Km/year x capta or 8 km/day x capta. In the USA, air travel is the most used for intercity and international travel and cars are used a lot in cities and urban buses are minimally used, but metros are widely used. In Brazil, air travel is little used, but there is strong growth in intercity travel, but few Brazilians travel internationally. Cars are increasingly used in cities, urban buses are widely used and there are still few metros in cities. In China, air travel is very little used for intercity travel, but it is growing strongly, although Chinese people are banned from travelling internationally. Low-speed trains are widely used for intercity journeys, but the greatest mobility is on foot and by bicycle. Urban public buses and metros are still few in number. Comparing the GDP/capita of each country shows the relationship between purchasing power and the distance travelled in these 1 hour and 15 minutes per day. In 2013 the GDP/capita was: USA 49,000 US$/capta, Brazil 11,900 US$/capta and China 8,500 US$/capta.

Another important factor in individual mobility is the average travelling speed. In the USA it is 46 kilometres per hour, in Brazil it is 18 kilometres per hour and in China it is 7 kilometres per hour. It can be seen that in the USA the difference is proportional to air travel. We can conclude that for everyday individual urban mobility, the average speed is very low and fully justifies low-energy mobility. By this I mean that the amount of energy spent to travel 1 kilometre should be as low as possible, or in other words, it should be as efficient as possible. It is known that at low speeds the energy expended is mainly proportional to the total mass of the driver and vehicle.

Analysing energy efficiency for urban mobility should be based on the energy expended in a normal walk of 1 km. A 70kg person uses 0.2 MJ/km of energy.[2e3] I'll take this figure as a reference for the maximum energy efficiency of individual mobility, i.e. 100 per cent. For example, if we compare the energy expended to travel 1 kilometre in the two most common types of car in Brazilian cities, a 1 litre people carrier and a 2 litre SUV, we get the following efficiencies. In order to use real market values, I took as a basis the data from a popular Renault Clio Authentique 3p 1L / 16 V with air conditioning and power steering with energy consumption of 1.58 MJ/km and a Hyunday Tucson 2L / 16V SUV with air conditioning and power steering with energy consumption of 2.75 MJ/km.[4] Compared to the energy of walking, the popular car used more than 87 per cent of the energy, i.e. it achieved a relative efficiency of 13 per cent, and the SUV used more than 93 per cent of the energy, i.e. it achieved a relative efficiency of 7 per cent. This analysis can be extended to all modes of urban transport, and thus reinforce the actions that the community must effectively take to improve the overall quality of urban mobility.

As a conclusion, I can cite the political considerations made by Schafer[1] for interested governments. A government policy can aim to rebalance the mobility preference of customers, induce a rapid switch to energy-efficient technology, with options such as: market regulation measures, urban areas with traffic restrictions for cars, and with criteria based on economic efficiency, impact on mobility time, customer acceptance, transparency and equity of mobility modes.


References:
1: Andreas Schafer, John B. Heywood, Henry D. Jacoby, and Ian A. Waitz - The MIT Press, Cambridge, Massachusetts. London, England.

2:http://sistemas.eeferp.usp.br/myron/arquivos/3396411/316f9dcf5e9599de79d7aa387aa6 3e6a.pdf;

3: Estimating the energy expenditure of walking. Leandro Nogueira Dutra, Vinicius Oliveira Damasceno, André Calil Silva,Jeferson Macedo Vianna, José Marques Novo Júnior and Jorge Roberto Perrout Lima. Rev Bras Med Esporte _ Vol. 13, N^ 5 - Sep /Oct, 2007

4: http://www.inmetro.gov.br/consumidor/tabelas pbe veicular.asp

# Vehicles per inhabitant according to Curitiba's neighbourhoods - 2010

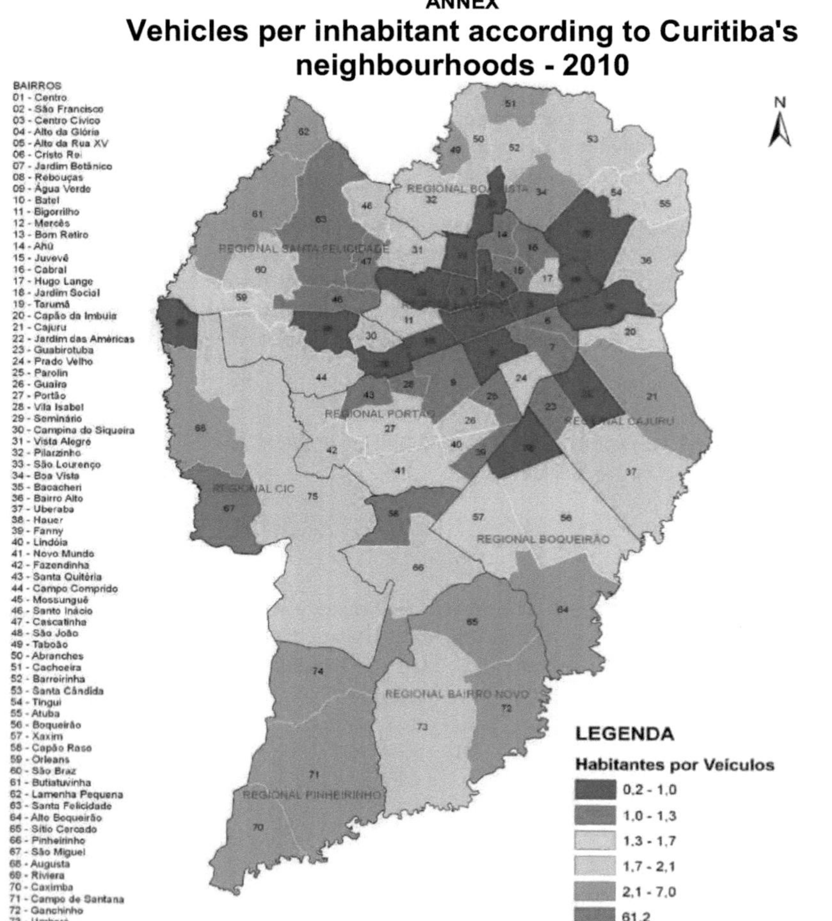


**IPPUC** Instituto de Pesquisa e Planejamento Urbano de Curitiba - SIN - Banco de Dados
:: Rua Bom Jesus, 669 :: Cabral :: Curitiba :: Paraná :: CEP 80035-010 :: Fone (41) 3250-1414 :: Fax (41) 3254-8661 :: E-Mail ippuc@ippuc.org.br ::

Printed by Books on Demand GmbH, Norderstedt / Germany